Finished Crea

new poetry

issue one spring 2019

Finished Creatures - issue one

---

first published in the United Kingdom in 2019 by Finished Creatures

a CIP catalogue record of this book is available from the British Library

ISBN 978-1-9160547-0-7

printed and bound by one digital, Brighton, UK using FSC certified paper

www.finishedcreatures.co.uk

Finished Creatures is an independent, no profit magazine, produced by skilled volunteers as a platform for emerging and experienced poets. We publish biannually in spring and autumn with two month-long submissions windows in January and July respectively.

Finished Creatures

new poetry

Airborne

Airborne

Precarious times call for more poetry: more ecopoetry, more political poetry, more poems of beauty and daring.

Welcome to the first issue of *Finished Creatures*. We put out this tentative call for submissions just four months ago, hoping for poems that expressed 'something new and restless'. The response was immediate and positive. Thank you to all the poets who took a leap of faith and sent us their work.

As we go to press the future feels even more precarious, with politics in disarray and policy-making in limbo. The overriding crisis of climate change – and the devastating implications for human and non-human life - is being sidelined. Conflict seems to prevail.

But poets continue to create community, speak thoughtfully and strive for better words, ideas and images to explore human experience and relationships with others on this planet.

Finished Creatures may sound like a pessimistic title, but it is inspired by Emily Dickinson's poem: 'The Chemical conviction That Nought be lost/ Enable in Disaster/ My fractured Trust.' The poem is reproduced on the inside cover with full permission.

This issue is Airborne: birds, sloths, bats and gods take their place on the upper branches or ride the thermals. Humans can be found in the mix: questioning their dominion and trying hard to make sense of it all. There is much beauty and plenty of daring.

Jan Heritage

April 2019

Contents

Contents

Contents

Jemma Borg

A quiet revolution
after Rebecca Solnit

We need a new story. And we will have one.
Though nothing is certain, that gives us hope —

we're yet to find our ending.
The storm lays a seed, and the seed must grow.

Among the blue shadows of the body,
the clock's two swords resume their wounding —

it's called risk, my friends, if you will join me.

Philip Gross

Of a Species that Falls

into flight, for the momentum
of the long low swoop-dip out of danger,

skimming the leaf mould
and riffle-grass (both unaccountably

not here, just tarmac) here's
one, in its moment: here's a million years

of outstretched evolution,
fanned out like a winning hand, to the tip

of each flight feather, trimmed to its niche
in the air – translucent

where the light strikes: an array of panic
and precision. This is it,

uplifted, not yet roadkill: the moment before
speed's fist, from no known

dimension, ours, a thing that passeth
understanding,
hit.

Matt Howard

Common Buzzard
19/1/16, near Kikinda

Legs not so much cut as wrenched off.
Yellowy splinters, bloody icicles
shorn in this freeze.
Milan holds it up;
its right wing's threadbare dangle,
flightless now:
a dreamcatcher stirring on no wind.

Crushed supraorbital of the dead, right eye.

I miss the fullness of what Milan is saying –
something about the luck of a bastard hunter
who either clipped him themselves or chanced on him.
Hit, most likely, by a high-sided pick-up
or an artic hauling from Belgrade.

They hand-in legs in return for cartridges.

But still that fixed scald of the left, live eye.

I've an idea of the structure
from diagrams in books or images
of raptor eyes online: a balled mass of rods,
cones; oiled receptors of the fovea,
honing through the optic nerve to the brain.

His beak gapes.
a cloudy eyelid
cleanses again;
with full acuity,

A side-swipe of
(nictitating membrane)
he is looking at us
our absolute proximity

at the verge-side as Milan ends it.

Katrina Naomi

Swaling on Boscathow
for David May

The farmer tosses purple moor grass into the air –
An Easterly – we're ready to burn.
Pale grey puffs become a bonfire guy legging it
through gorse and bracken. The fire doesn't burn
into the ground, is all surface and speed.
I'm responsible for my patch but if the wind were to run
away, I'd sprint this grassy corridor
or be sacrificed to some god I've yet to hear of.

Annie asks me what to do
as if I've worked a fire this size before.
The contractor, who set the fire,
would be handsome on a film set. Here,
he's in frayed red t-shirt, hole-ridden overalls
open to the groin. I like that he says *he*
instead of the English-English *him* –
I've worked with he before.

The fire finds its own voice, stalls,
worries at a tussock before lurching on.
A shrew bolts. The fire leaves
part of itself, teasy as an orange baby,
writhing by my boots. I do my job and whack it.
The farmer nods – *almost done.*
We breathe in the distant danger,
paraffin still rides the air.

Claire Dyer

Spill

I Study For

Mackerel are shoaling the blue waters.
There's a docker dividing a clamour of human hearts

into wire baskets on the quay.
I watch the growth knots on the clematis in my garden loosen.

In the tiny are vast, inexplicable things.
The first thoughts will be of birdsong captured in a jar.

With practice I'll write the splitting of glass,
the spill of the song.

II Spill

She lifts the jar with her pale hands,
her pale fingers,
her fingernails mother-of-pearl pale.

She twists the jar in an act of refraction
to the point where light splits,
waits –

waits for the heat to build,
the first cracks to appear.
The sun is at its absolute.

And then they come: hesitant, sinewy,
like moments before deer run,
and soon the sound is deafening –

the foretaste of apocalypse.
So the glass shatters, blasts tiny
red stitches into the skin

on her arms, on her face,
yet still the glass glints,
yet still the light splits.

And, at last, the song spills
in a song waterfall.
The birds know exactly what they do.

Jane Wilkinson

Cormorant maps

each day   a new document

each day like a bullet

committed

to cormorant's lists

it navigates like this

this

footpath this road this factory   common

this motorway   this   green field   this

marsh dump   this   wind turbine this water tower this   sea

these remote wings baling out prehistoric shadow

under the wind   compressing fish guts to jet muscle

I look

at the cormorant turning

slightly

making a correction

like a single punctuation mark passing by

with a fisher's eye

it threshes the line from the light

silently over the screen

as it passes by

a comma   comma   sky

Paul Stephenson

The Die-off

Extreme heat wiped out almost one third of Australia's spectacled flying fox population.
(ABC News, 19 December 2018)

The ecologist collated the large numbers,
took stock of the freak occurrence in Cairns –

almost biblical, the colonies of flying foxes
dropping heat-stressed from roosting trees.

Local volunteers used garden wheelbarrows,
shovelled up the carpet of rotting carcasses,

smelt the stench. Early in the birthing season,
eight hundred and fifty rescued, the orphans

laid out row after row, each one tucked tight
in a wrap of white towelling, like a bat burrito.

The mercury soared one afternoon: 42 Celsius,
November's record broken by five full degrees.

She said, *None of our carers were prepared*
for the numbers. We went in flying blind.

Pollinating

who knows of this grace:
grown into the nucleus of meaning:

that courses through us:
along genetic streams:

and is often forgotten:
until the point of extinction:

leaving shapes and silhouettes:
fossil language:

flowed and twined:
to symmetry to cemetery to:

Lisa Kelly

Clutch

Slapped away his ovum-filching fingers

I am not cuckoo
I keep my eggs about me
clutched and warm-nested

Climbed a branch and brooded

Deciduous girls
dropping eggs like leaves
yolking the dark earth

Eggs eyed-up and guarded

A golden goose kept
by a medical giant
lays golden gametes

In the Forest of Anagram, a warning: Fire is Rife!

A fibrous nest burns
thin twigs up in smoke
as the moss smoulders

Eggers, oologists, experts plan raids in the spring
On Holderness and beyond

How much more thieving?
Peacock eyes watch eggs blown through
in a begging bowl

Suzannah Evans

The Recipients
After Jon Lomburg and Carl Sagan

One does not meet oneself until one catches the reflection from an eye other than human. — Loren Eiseley

The recipients will know
that dogs are our friends

that we are not just runners
but spectators of sports

they will deduce
that we catch animals and cook them

although the mouth is complicated
we have demonstrated eating, licking and drinking

they may or may not understand
an economy of buying and selling

their eyes may not use
the same spectrum of visible light

they may not have senses
in the way we experience them

they may already know that all life
is made up of DNA

They will recognise earrings
as decoration, we hope
rather than miniature radios

will they realise the sunset is beautiful?

we think they could figure out
that a vibrating string
makes a certain kind of sound

they will not see the UN building as symbolic

Freeing stuck vehicles
is an experience they may share

They will see that our planet
has conceptual borders marked by lines

though they will learn the texture
of our star

and gain an understanding
of the volume of the human form in space

they will not see the worst of us
that more have starved to death

or died in wars
than have written symphonies

Bad Martin Takes a Hostage

If you're anything like me, then
you'll hate the idea of being
suddenly upside down, change
raining down on the concrete,
your soaked blue jacket clung
to your face while you swing
head heavy with blood in the dark
wondering what happens next.

Susannah Hart

Plague

Yesterday the authorities put down her pet dog
to try to stop the infection spreading. He was a good
dog who always came at the sound of his name,
a thin brown dog with gravy eyes. She was looking
after Spanish missionaries who had built their confessionals
in small hot towns spread by blood, sweat and saliva.
The dog's tail was prone to wagging. He had a rubber bone.

Others under observation include two hairdressers
three medical workers one forklift truck operator
one pet shop owner. Rapid construction of facilities
is still ongoing and extra screenings have been arranged
on giant monitors at all major cinematographic venues.
Vets are wondering if mass euthanasia will be required.
The dog liked to eat chicken and was usually safe off the leash.

The Madrid suburbs continue to pose a biological risk
and there is evidence that more dogs will come to heel
in the near future, across the parks and plazas of the city.
Travellers are urged to consult their guidebooks which may be
oddly silent on the source of the contagion. They are further
advised to follow all protocols regarding the stroking of canines.
The dog used to bark at strangers but never for long.

He gives her directions

From the ruined hamlet, strike out over fields. You'll become aware of a sewage works at the first T-junction, turn left and continue north; ignore the level crossing straight ahead. When at last you see the lake, board a ferry near Foxcote Farm. Coldblow lies south of East Stanton: its tumuli are vast. Note three gravel pits stocked with giant carp. Don't be distracted by a crimson road — some days the gradient is more than one in five. Bell Avenue usually leads from the canal to Myrtle Hollow where you may see a disused pumping station, Norman keep and crematorium (with tower). Stand with your back to the row of blasted elms.

Jane Lovell

kestrel

wired
to weave
perfect light

wind-buffeted
winnowing ocean-air
turbulence

pharaoh-eyed indigo-deep
resting
momentarily
in dark-path leaf-curl
scent-seeking
nightjar-crouch

then
eye strike
shadow-patter
grass-spire quiver

cryptic tilting
dipping
sudden
tunnel-thrust and

lift
lift
adjust

accommodate
the
crosswind

hover

hold

hold

hold

my beautiful

kohl-eyed

tempest
angel

if I'd gathered enough or held you

a mermaid's purse lightly attached to weed
maybe there would have been no cramping
or tides bearing away the floating foam

or if I'd fallen asleep sitting up maybe
I could have clutched you safe despite
a precarious nest on the basalt ledge

maybe if I'd been a cave swiftlet building
a saliva-cup of stringy vegetation sticking
secure to walls if we hadn't been prey

maybe if there'd been no buffeting
no white shock if I'd had the strength
pale fingernails clawing rock or if crying

I'd have caught you before the fall maybe
I'd have felt your heart or seen your eyes
instead of black wings beating

maybe if I'd been no crack or
salt-crusted shell releasing red
if I'd been enough love maybe

Sarah Doyle

A Song of Asking

We asked the bird why it would not nest, why
it would not nest. The bird told us to ask
the branch, and so we asked the branch.

We asked the branch why it would not bud, why
it would not bud. The branch told us to ask
the root, and so we asked the root.

We asked the root why it would not drink, why
it would not drink. The root told us to ask
the rain, and so we asked the rain.

We asked the rain why it would not quench, why
it would not quench. The rain told us to ask
the earth, and so we asked the earth.

We asked the earth why the bird won't nest, why
the branch won't bud. The earth told us to ask
ourselves, so we asked the earth again.

We asked the earth why the root won't drink, why
the rain won't quench. The earth told us to ask
ourselves, but we only asked the birds.

Early flight

In the aquarium of dawn
constellations of fish
with orange-lamplit skins
and spiralling limbs
swim either side the estuary.

Skate, electric eel, ray
keep their distances,
tails thrashing, mites
streaming along the channels,
no end and no beginning
endlessly carrying tribute.

Swarms of moon jellyfish
form clouds, part, reveal
exoskeletons of light.

From each of our fish tanks
we watch one another,
each one thinking
we see the whole.

Anuja Ghimire

My American Child

I love America like I loved bed
rest for five months
for my unborn

if every worry is a contraction, and I need
you whole baby, I practice
breathing the way you will
learn fire and shooting drill

I came here to be somebody
—but mostly yours

fortuneteller speaks of your house by the sea
and I hang on by a straw

my landlocked memories learn to love your soil
it's reddish brown

when you arrive in winter, I'm gaining atrophied muscles
I am warm like my blood in you

I name you, raise you above the sun, and watch you carry the light

Judith Cair

Chantry Hill

The petals of wild
marjoram have fallen, their stalks
stud the Downs, maroon bracts
like tiny second flowers condensing
September sun. I move from
hummock to hummock of
Cross Dyke, watching
the land fall away into the
valley, soar up to the iron
age fort. And I am
weightless, just as in my
dream, I who have
neither painter's eye nor
bird's, but sometimes in my
sleep am given both.

Cheryl Moskowitz

Hummingbird

I'll start with firsts since I can't do endings   I'll need a new hummingbird/ticker tape/whistle pipe/gumbag   there are no word - murmurings   yes that will do   let's start with those   soft mmmmms mumma mumble mother [that's sweet but not how it was] nothing back-boned or hard   Nothing   no one/swirls and some shadows yes those quiet   see?   I can say the word quiet and mean it/I can know this emptiness before something   alone when alone was pure... sweetness all treacle and amber   whatsitcalled wassicalled? without consonants sibilants mouth pops just inside yourself   there is no *you* inside   too earnest   too hard   inside *her*self   yes that will do   no hims   cease hymning only soft sibilant esses   ok   yes   we'll let the sibilants in here shall we?   at this point of beginning   all import importance squelch not even formed yet   a tadpole   only that's not nice not even right   too... wasstheword   clichéd   just a nothing swimming in something   wet/dark   but how would I know?   colours don't figure in ~~this place~~ here   they don't we had all that[no you yet but there was a we/an us/a them]   had to be else/how else am I   is this here at the start the beginning   need a new hummingbird or something tickertape   maybe a tree or a shadow of a tree   only I don't know these things yet   not the thing or its shadow only felt   yes feeling's good   I know that much   there is this feeling   a sense   warm even   a sense   a surround   yess yes surround works   we'll have that   we're getting somewhere   sssssssspace   sssssssilence   ssssssurround a proper beginning

Cornflowers

Autumn is near. The cornflowers are fading.
Their pink and blue coruscation diminishes
slowly, but is still enough to attract attention.

Those wild seeds were planted here in such
a deliberate manner, to muster some colour
along this wide expanse of municipal grass.

That playground is full of rattling children.
Noise of their singing and games is muffled
by humidity, smothering sound and movement.

The day is winding down so, when the storm
begins to rumble, families depart for home,
climbing from one flood plain up to the next.

Lesley Sharpe

Surprised! Tiger in a Tropical Storm, after Rousseau

Cayahuari Yacu, the jungle Indians, call this country 'the land where God did not finish Creation'. Only after man has disappeared, they believe, will He return to finish His work. Werner Herzog, *Fitzcarraldo*

To be honest, I wasn't surprised.
Knew it was coming, knew I would know
the predator when I met him, being so like me

in his restless hunger, though he didn't know
when to stop. When rain fell from the sky –
isn't that where storms always come from, splitting my mind

with electric force? - at first it was just a breeze,
those wild drops, then the full awful catastrophe.
For some reason the density of foliage made me feel

safe, the thick green of it, leaf upon leaf, wild cut-outs.
I'm surprised, I said, *that you've let it go this far* –
now this is the only patch left - *but let's pose*

for the camera. I bared my teeth, pawed those leaves,
got a bit tangled, flipped up my tail, *a striped snake*
in the grass, he said. I knew that thin wire of electricity

would light the scene for a few seconds, a brief spasm
before my flame red flowers would go out, leave
only his name, in huge dark letters in the corner

- his picture. But this is my jungle, remember?
What he can't see is still under my paw – ants, tarantulas,
and those blue butterflies bigger than the span of his mind.

The sky flashed. *Let's rehearse it again,* I said. *Quick!*
But remember - you are the savage, though for now, I'll pretend

Thief

after Simon Armitage

If you could eat larks' eggs
you might think they'd taste sweet,

the food of the gods,
that melts on your tongue

so that, afterwards, you'd sing
the clear liquid notes of the bird

whose nest you had robbed.
Not so. They'd stick in your throat

so that all night you would taste
the shocking gold, the bitter white.

Jemma Borg

Portrait of shingle and wildflower

Stones the colour of bruises, cabochons of white water,
stones with an unreadable text
– scratch on scratch of ratcheting tide –
stones turning to sand, sand turning
to stone, down the millennia
in chuckle, clap, block and tumble, knock on knock
and rustle:
granite stones, flint stones, brown stones, blue,
the water sieves them all
and spits them out and grinds up
crab legs, whistle-empty
shells and brittle stars,
and the small stones are morose, shining
with silver scars, indented with greys
like waves bobbing sometimes
with momentary blacknesses
(or is it a trick of the eye?),
then building terrace on terrace
now ebbing, now fettered again
with sea kale, a succession of stonecrops
and bittersweet above the strandline
where they settle, fall,
shift
and settle –
may even gather soil –
before the autumn storms re- open them,
submerge the ridges and
shake loose the wild-
flower seeds:
the unbound gems of a bound life

Hannah Rose Woods

Neomodern International

The peregrine falcon is a cosmopolitan bird of prey.
They sometimes nest in the A's of Arndale Centres,
and also in New York, I believe, chasing pigeons into
skyscrapers. The other day I saw a picture on Twitter
of falcons on a plane in the United Arab Emirates. There
they were, sitting in first class in their little hats. So fancy!

Claire Collison

Trigonometry

If I am A, and the plummeting paraglider is B,
then José María is X—
the third angle, the still axis.

If I could show you a ground plan, or aerial photograph—
José María has one, taped to his wall, taken,
as it happens, by the paraglider.

José María is *front line:* there is nothing
between him and the sea—when the moon lifts,
it throws its silver directly to him.

My apartment is further back: José María
interrupts my sea view by perhaps
twenty per cent. Others impede

the mountains—a view José María commands—
but he semaphores, advising me
of impending weather.

I covet his terrace, which is capacious.
He paces it like a ship's deck.
That's where he was when the paraglider

dropped from the sky, tangled in pylons, we learned.
I tell you, because this is a story of perspective:
when the paraglider fell, I saw only José María's response—

he uttered a moan from his lonely terrace.
All the pettiness of neighbours' dogs, etc.
fell from him.

I heard the sirens. I saw the ambulance.

Air Traffic Control

The planes are inconceivably
pinned overhead by grace and terror.
A sudden loss of radar's left
their pilots bare of navigational support
so they must plot their downward drift
by gaslight, Venus, almanacs and gods.
Allow them if you must to choose
their own devised machinery.
Use all your forced imagination
to will them safely home.

Michelle Penn

reasonable doubt

everything depends on frame of reference
        not events
                but the relationships between

gravity wraps itself around mass
        space kinks time kinks light

                        and here, in this house
        of higher law, wisdom feels tenuous

classical physics can chart a bullet's path to a body
        but what can it say about singularities

or why time stops
        for you but not for me

everything is relative:
        the pup-eyed defendant in the dock
and if doubt is reasonable

          all we know is maybe he did it

we've been called to judge one side's story
        more credible, one set of memories

less creased, to remain
        unbound by emotion or experience

                        reach toward the edge
of uncertainty, where the plausible
                        seems absolute enough

if Einstein was right, you always perceive
        one dimension fewer
                        than you're in

Café Gijón, Madrid

I've been with dust and pigeons.
The sun has touched me.

In the cool of this panelled room
I miss you.

All through our year of grief
you couldn't warm to this city —

The streets are empty, you said,
the buildings too grand.

The columns inside the café
don't seem to hold up the ceiling.

If you were with me, we'd peer
into mirrors, interpret the signs.

Charlotte Gann

My Brother's Long Journey

His skin itches like a coat
he can't scrape off.

His heart pumps doggedly.
Yanking his shirt on –

over all this – is difficult.
One sleeve catches

and he yearns to rip his
nuisance dolls' arm off.

End the misery in blood.
He tackles his buttons –

fucking fiddly things, plus
two at the veins on

his wrists. He stares in
the mirror. Back at him,

hard-eyed, comes that look.
His glasses glisten. He

needs to be *held.*
The bathroom floor is spinning.

His bag stands in the hall.
Where is the fucking taxi?

Paul Stephenson

The Hot Air Balloons of Finsbury Park

after Annabel Grey's mosaics on the platforms of the Piccadilly Line, commissioned by London Transport in 1983, their 52 colours of tiles shipped from Vincenza. 'On 15 September 1784, Vincenzo Lunardi became the first human to fly in England, accompanied by a confused dog, a puzzled cat and a seen-it-all-before pigeon' (londonist.com). The balloons actually took off from Finsbury Fields near Moorgate.

See the six with their backs to the wall,
all hot-air and helium with some dream
of unpeeling – detaching load tapes,
loosening tethers without rip or snag,
to float off silent, unnoticed and unmanned,
into a late-night tunnel on the Underground.

Cobalt cloudhopper, she'll take the lead,
her ruby eyes piercing a soot-dark path,
snatching taut the ornate gores running
up from her striped throat to crown ring,
her double seams lapping down the skull cap,
her forked banner a show of zigzag stitching.

See her gondola, a round-bottomed galleon
not woven in wicker or rattan, how it's adorned
with opal and onyx. This phoenix's bulb
is silver and gold, its silk ribbons curling out
from plain tiles, her pressure vessels gimbal-
mounted, tanked up, awaiting buoyancy.

One day, too porous, and too worn to fly,
this giant air-reliant might be retired, used
as a rag-bag, cold-filled by whisper burners –
resigned as a puffed-out cloud for screaming
kids to run through, her silicone rip-stop
nylon some vintage sag, all degraded coating.

But today, hey, she's invincible! impermeable!,
a pied-piper heading south, all five following
firmly behind – off to explore Turnham Green
possibilities. Or if the wind changes, she'll take
the Victoria line and propane north, past duelling
pistols, convoy up to Seven Sisters, hail hotspurs

of Tottenham, the gang gust-galloping, each in tow
below the Blackhorse Road – then exit the sky blue
end of the line, where she'll ramp up the heat
to carry them east, across the continent to Anatolia,
and marvel in the glory of fairy chimneys, six balloons
on a cruise, high on the plateaux of Cappadocia.

Woman's Work

hand stitching orpinment
orange leaves as they fall
from a wind danced branch
that all but touches the lake

sky gazing ochre wing feathers
of a kestrel beating beating
in search of prey in fields full
of sienna yellow lady's bedstraw

breathing the distant scent
of ripened wheat and barley
on long dry August afternoons
while days draw in to darken

colour is light and light is colour
honey yellow to straw yellow
wine yellow to gamboge yellow
then a dash of chestnut brown

James Goodman

The winds are up

all the winds are up all the gulls are up

all the gulls are up and can't get down

all the trees are up in Mellingoose Woods

crack! all the pheasants are up

and the guns are up

the sweet slurry smell is up

up in the air with its drunken gnats

and hey look!

the rooks are in the air

like a basket of laundry

the quick water of the stream is up in the air

the curlew's flute song is in the air far off

the clouds are up commodiously rearranging

and we are up too grazing and crazing around

and the land has no purchase

Marion Tracy

The Cliff

At first meeting, shock of the vertical garden.
In the moment of her entrance,

silence makes the air seem larger.
A list of gestures, body parts are shown

in slow motion. He takes a step too far
forwards, a heart wakes, a head turns.

Chalk is the colour of distance, what breathed,
what hung in the air between them.

There's wire netting over the brane
ready to catch what might tip down.

At the cliff's face, their startled shadows
fold and fall, drifting together across the light.

Pippa Little

Battleship Wharf

A spit of land between waters estuarine and salt
where Soviet hulks half-broken like wild horses
brood in the wind, the spiked north-easterly
which slams me sideways, stings my mouth
as if I've swallowed blood
or licked red rivets

and nothing changes here
in its brokenness
except grass, salt grey, insists on its community
among cracked stone or concrete,
three silos empty and fill with grain
according to the moon:

nobody comes for us, surfaces wear thin, walls fall away,
from high above, seen from a jet or drone
nothing is written in the thin snow
of my childhood's footprints

Becky Varley-Winter

Eden (full blown)

I.

When the storm rolled away,
when the storm left us,
when we were in the wake,
when the leaves hushed,
when the mouse rushed on her fugitive belly,
when the apples were pinned with dew,
when the small oranges waited to drop,
when the captivator berries did too,
when petals glowed like lit-up blood, lying in the limelight,
the magpies croaking an unknown language
when the river flooded,
when the whole garden looked like paintbrushes,
when I saw the edge in your eyes,
when you led me outside

II.

Not knowing, see a gang of crows
clamouring between two trees
a cloud waiting with puffed cheeks
as starlings perch in static air, chattering
feet clenched on the nerves of the pylon,
an electric snake. Every drop of rain
scatters into small applause
and when I trace your face
it thrums with a force
I can't feel in my own skin.
A question-shaped balloon
just floated past our window, glinting
as parakeets flock, apple-green
through the eyelines of the hawks
hunting far around us.
Arrayed in light
fractal mistakes, how long can we stay?
I've seen the first blue butterflies of May
and many full-blown yellow roses.

Tim Youngs

Amphitheatre

Not shadows of the past, then,
but of birds that zag between
the ocean and the old town,
over lovers taking selfies
in the area for wild beasts.

On Mussolini's steps a girl
jumps repeatedly, her legs bent
in triumph at each highest point.

Outside the north-east wall I pass
a woman who wants none of this.
Instead she stoops to photograph
a feather by a paving slab.

Joel Scarfe

April 1982

Out by the back door
the cat had miscarried her batch
and was licking clean
the syrupy peach-half bodies
when my mother shooed her away
so she could scoop them
into the bin.

I was six years old.

The radio was full
of an abhorrent song of conflict –
some distant islands poised
to receive their dead.

Something must have happened in my head
because, now, much to my own
children's distress
I will not have animals
in the house.

And as for armed hostility
I tell them
don't even think about it

not under my roof.

Robin Houghton

Breath

She murdered her first two husbands the way she knew best:
a bullet list of lies, a game of gaslight.

Now her daughters say she flies *up there* - somewhere unspecified
with angels, watching, sending blessings out of death.

But when a chill wind scolds my cheek in March   killing Spring
curling leaves and drying new shoots brown

I know it is her brittle heart behind it all, it is her breath.

Katrina Naomi

Interpretation

With thanks to Yamanashi Prefectural Library

I've almost stopped interpreting
yen – all those noughts.
I thought, at first, those notes haven't helped me
write a poem. I recalled a man on Waterloo Bridge
who wrote poems for cash. I offered £2,
received a poem about love on orange paper,
a purple envelope. I could have paid £20, £20,000.
What could a poem be worth and to whom?
How many noughts should I add,
for a favourite poem? And how could it be owned,
no matter how many yen or pounds
in a shiny gold purse? None of these
philosophers in their remarkable robes
can buy such words. A poem's worth
everything and nothing. Perhaps
some of my philosophers understand. And yes,
it has cost me to sit in silence
in this spacious, air-conditioned space,
the philosophers asleep in the close confines
of their dreams. What would Austen say
on the matter? And if I threw these notes,
these dreaming philosophers, from the top
of this building, with its roofline trees,
typhoonish-blue sky, who's to say –
from such a distance – what is money,
who is royalty, what are mere jottings,
and which is a love poem written to a stranger?

Claire Collison

Dolomites

Especially hard hit were the spruce trees of the Paneveggio park, beloved by Antonio Stradivari as the place to find wood for his stringed instruments. (The Times, Nov 2, 2018)

I almost blinded a cellist once, in St Martin-in-the-Fields. He was performing the Bach Suites from memory. Sunlight ricocheted from my pocket mirror, just missing his eye. When I dream I am flying, it is modest and domestic: I take off from a chair, breaststroking over dining tables and kitchen work surfaces. Omniscience has always been cherry pickers, the first five minutes of Citizen Kane. Drones give feeble God-view, but we can't resist. How unevolved we are, compared to flies, or root hair! There are cellos in the trees. We wear paper hats and complain like squirrels, feeding our leftovers to the wolves.

we can always find each other
we sloths with secrets

a diet of leaves falters
time's engine
muscles lean across
moments of stillness

i wear my edges
a scaffold of expertise
& exercise
as visible rebuke

unable to
shiver when cold

teach me how not to need

fat deposits
only on feet

bee-killer bud-witherer
i may not see the sun
nor the sun see me

teeth
high-crowned
open-rooted

hair grows grooved
thick
with algae
strange
joints

virtually no external
sexual dimorphism
i avoid being eaten by eagles

i cannot properly see
i run from my own image
flesh & feeling

i have to be
everything
hanging on
by my claws

my mind's glass
struggles
against this
invincible life

sewn up
in a hammock

in a dark narrow
cage

suspended between
heaven & earth

my soul escapes as a white mouse

i will try again tomorrow

float across tree tops
drink out of a swan's bone

from The IKEA Back Catalogue

TÄRNAN

1. Common

I turn on the tap of my tern
and water dives in the sink,
the egg of soap precarious
in the scrape, the small
depression in the ledge
overlooking a sea of plastic plates.

2. White

I turn on the tap of my tern,
imagine a tropical isle,
the egg of soap in the fork
of my thighs branching out
to float in the infinity pool
flowing over the negative edge.

3. Arctic

I turn on the tap of my tern,
a single lever chrome faucet,
an outstretched grey wing
holding out for the Arctic
the Antarctic, spanning the
globe, its mission circumpolar.

Suzannah Evans

What is it Like to be a Bat?
after Thomas Nagel

For a bat to be a bat, I mean,
to use its whole body as an organ of sense

to rattle through the high-pitched dusk
feeling the geometry of cave walls

crunch the exoskeletons of moths
and taste their powdered wings

to sip in flight from the surface of a river
to fall and elbow itself up off the ground

then ping back into the air
like a rubber band

to blanket itself with its arms and hang
from its feet and not let go

to slow its heartbeat into winter, wake
with the hunger of a season's sleep

to tangle with humans in the lofts
of old buildings, feel them lumbering

slow as planets through space
to zip between their heads

gone
long before the gasp.

Matt Howard

---

faire le pied de grue
Slano Kopovo

I want to tell you of the slow flatland miles
to that frozen salt lake, barely a soul
out there, whether waiting or in passing
while the frost-blue sky is as near pristine;
almost cloudless. Worked miles with little relief,
just stilled rivulets of roe deer, golden jackal
tracks and tractor runs through alfalfa
to the horizon, where the first cranes come
low, out of nowhere, two adults and two young,
seemingly going nowhere, unlike anything
you think you've seen, until you too arrive
at Slano Kopovo where they stream in
from every direction, sudden thousands
bugling their racket and heat before
settling on the iced-over shallows.

Radiolarian, by Blashka

Aulosphaera elegantissima Haeckel
(glass model, National Museum of Wales)

Sometimes it takes such craft,
such curiosity,
to come near
to one life, in the lightless
weightless depth

or height that holds it
perfectly, its floating
moment, its animate
star.
It might take glass

that nature gives us
nowhere except look,
in a season of storms
with thunder wandering
across bare desert,

in beadlets of silica
fused to black glass
by a lightning strike —
as close as we can picture
to the tip

on tip of Adam's
on God's finger or
the flash of pure attention,
there, wherever
there is. Absolutely there.

Asim Khan

Dendrochronology

stone orchards' secret(e)

before the yawning creaks

. . . swirling in false-light . . .

perched

the passer refrains:

[the passeriformes]

echo to fossil . . . //. . . //

Alastair Zangs

Axis

Seven-day working weeks
and what to show, but

one sorry square in a field of nine;
abandoned noughts and crosses.

Wheat can't self-sheaf,
so when he died on harvest moon,

and the shot rang around
the lonely acres,

that golden parcel sat pristine
beside the idling combine.

Instead of a twenty-one gun salute,
they lined up a bird scarer

to throw its compact pops
across the way. It's all bluster though –

cocky beaks pull up green
shoots as the cannon carries on.

Scattered grains fizz across
the flat roof of his wooden box,

placed in the ground where
the sugar beet should be.

Matthew Stewart

Under my Breath

While ironing, shaving,
pouring a cup of tea
and driving home from work,

I catch myself singing
the same tunes or jingles
or terrace chants as Dad,

imitating his tone,
occupying his space,
adopting his cadence.

Claire Dyer

Virtual Swimming

When water got scarce, still
she'd rise at seven, strap on headgear,

load the app, remember old mornings:
wave slice – head down, breathe out,

head up, breathe in; sunlight
disco-balling the blue,

skin sheening with the wet.
Remember counting laps,

pausing at the turn, nodding
to the guy in the next lane,

music seeping through the speakers,
poolside. Remember when it rained

(remember airborne rain?), rain dropping
like bullets, voices from the jacuzzi

a distant thundering. Remember
emptying her mind of all but heart pump,

muscle stretch, the cold jet a startling
in her blood. Remember the sea,

before the cordons came,
her pushing against the surf,

the endless drawing in and out of it.
Sometimes in high summer

when she waited for her breath to ease,
she tasted salt spray on her lips,

watched the tiles dance at her feet,
not knowing how soon it all would end.

Audrey Ardern-Jones

The traffic is lessening

now that lights in the city are dimmed.
A waiter fills his bins with leftovers:
chicken bones, spills of béarnaise sauce.
Another night shift, haze obscures the stars

and walking to the bus stop I hear
Sister Mary Dominica's clear-water voice:
Men, they nab you in the dark.
I avoid cracks on the pavements.

Stepping into Bladerunner, air

Monday, and the meeting will be long
and difficult, hot air, and I'll spend time
exploring which side I'm actually on

– now let me take a moment for that thought –

and I'll develop a headache and gulp a glass
of unbranded water before stepping
into the yellow air of early afternoon,

air heavy with dark, and as I dodge down
Oxford Street in a thick wind, air's sand
and ash will make a hearth in my mouth,

and the roof of the world will fall on me,
and the bus I'll take now will be half-full of air,
leaving me infinitely, without using up data,

to consider the left eyes of my companions
as they strain at the red sun, and
I'll breathe desert and burnt forest air

for as long as it takes to read this line.

Lilac

Born of sunlight, rich earth,
rain-soaked roots, this cream

sapwood, the hidden purple
heartwood, these handfuls

of pastel florets, their scent
sweet, but not too sweet,

petals radiant in early heat.
In folk law, to cut a lilac stem in flower

and bring it to the house
invites bad luck. This morning

I can't resist, so clip a bunch
and float it in a drinking glass.

So far, no harm has come
to us; the tap water

runs clear, the toilet
will still flush, and in the sky

two war planes pass
on their way to somewhere else.

Richard Price

The air that he breathes

I have a little boy,
late gift in last days.
He laughs so freely, and that's how he plays.
He doesn't see nothing's free –
least, not the air that he breathes.

I walk with him.
I take his sticky hand.
We risk the road,
he skips to a scrap of land.
Beneath old trees refugees twitch in their sleep.
We're all 'sharing the peace' – and the air that he breathes.

There's a five-a-side field,
it's all marked down for shops and flats –
'affordable' homes, and zero rate of tax
for land-bank owner-absentees.
It's a Government decree.
There's a short-term lease on the air that he breathes.

I never thought I'd leave this world
with the children fighting for air.
I never thought I'd see this greed
and leave them choking there –
outside, at the 'gated community' wall.
I didn't think at all –
I believe the things we need should be free,
including the air that he breathes.

I have a little boy,
late gift in last days.
He laughs so freely, and that's how he plays.

Specimen

Faded to a watermark,
I am best viewed held up to the light

kept behind protective glass
and handled only
by white glove-wearers.

A case-study in acceptable hate,
I will not be restored.

Katherine McMahon

Incantation

I cut my forearm
on the Shard;
let the blood fall
into my garden.
I take a seed I found
in the cemetery-turned-woodland
with its uncountable textures
and clearings floored in leaves
turning to humus turning to mushrooms:
I take that seed and plant it
in the clotted earth.

I breathe in outbreath CO2
on the crowded tube
like leaves. I know my cells
are floating through the tunnels.

I eat what I am given:
the canal-side blackberries
enclosing pollution that drifted
into flowers like stardust.
Know that you are not a cancer,
they say. Know that you are a cell
which is plotting to overhaul the city
by becoming it.
Let the black snot and smog cough
be your communion.

I speak to the thin fox regarding me
from the garden wall. She is not afraid.
She says, *this is land.*
This is what we have made of it.
I tell her, I cannot bring back the holy grove
that London was named for.
But I can be it, I can take my hands
and make them into trees.

Jean Hall

Conjugation

I make this prediction
you will figure it out
he was always a man with a mission
she was crushed by his dirty tricks
it is all about revenge
we must convince him of an epiphany
you saw the signs when his advisors jumped ship
they were replaced by ideological warmongers.

I saw his face at the Correspondents' dinner
you heard him say he'll find ways to win a war
he thinks that's why he was voted in
she helps him with his fusillade of tweets
it is not enough to say he is mad
we could witness the first apocalypse out of spite
you know what a pre-emptive strike means
they must stop the ticking clock

Lydia Unsworth

He Nuzzled Like a Dog Against My Calf

We're flying and we have forgotten to notice. A bird is caught in the propelling propellor. Bird strikes are a significant threat to flight safety. Says Mama Bird.

We are flying for love, for a better future. Our children are sulking but they'll adapt, they're young. Of my two sons I took the youngest first, despite his shorter wait, because he had more of a chance of not remembering. When they are older they'll fly home again, to that other place to which they also don't belong.

It has been estimated that there is one accident resulting in human death for every one billion flying hours. The majority of bird strikes cause little damage to the aircraft, however the collision is usually fatal to the bird(s) involved.

We do not like the food, we do not like the men, we do not like the weather. The nights are dark and long. We joke in our own language while the men hover by the drinks cabinets and try not to hurt us or themselves.

Do we want to be alone?

In addition to property damage, collisions between man-made structures and conveyances and birds are a contributing factor to the worldwide decline of many avian species.

I want humidity and sweat; hot and equal nights.

Years later I do not notice the cabin pressure, no longer pay heed to the safety ritual. The child has left to be anonymous in yet a third location as we, the guardians, stand about bickering over which one of us gets to die in our own world.

Caroline Hammond

Deep Water Warning

Once you've been into the river
it doesn't want to let you go.

Eyes check the roll and speed
of gulls riding downstream

and you feel the two bones
in your forearms press close together,

instead of breaking, as rescuers
pull you back into the air.

If you've been into the river
you notice what brushes your face

on its way down to the water -
paper, plastic, dead leaves, snow.

Hannah Rose Woods

Syracuse

On my not-quite-private expedition, past the resistance of water,
down to the darker place, where the detail was waiting—
I kissed your lips so softly,
mouth closer,
and the tide took me over—

I thought of you as the plane took off,
rumbling forward over dark waters and pinprick
light, bouncing lightly over the cloud line lightly turbulent,
bounding in low swoops triumphant with the thought of you—

I would have dashed the plane to pieces, playfully—
tumbled headlong into velvet night exciting, tumbled
forward into velvet night exciting, *oh, you—*

lightly opiated, I prayed to the ancient
and appalling gods, like Odysseus heading forwards
onto endless seas just forwards *(let me move, and do—)*
and the gods delighted, laughingly—

said this is not how it works, asking—
said I was travelling solo, anachronistically—
said have you *really* not worked out yet
how this ends for you? eponymous flower, best case
scenario—said I could set the earth on fire, soon as have you—

Jill Abram

Outback Nights

I miss the curlews calling
like ladies screaming.
It makes you think there's evil
out there in the night.

On fly-screen covered windows
the cat leaps at geckos
poised to catch the insects
drawn by the light.

Far in the distance
a rainless storm with lightning
flashes, jagged dashes
and the sky is briefly bright.

Fiona Larkin

Cohabitation: a remix

Neither self-sufficient nor intact,
we open up: offer a tacit

invitation to our body's habitat.
We dangle like a rind of bacon

from a makeshift line, bait
for microbe populace, titanic

in their numbers. It's both
reciprocal, like gin & tonic

and interlocked: the botanic
garden of our guts a chain

of nourishment in action.
Host and guests obtain

equilibrium, and we contain
. . . our multitudes. See us attain

through symbiotic chit-chat,
or call-responsive chant

within us, our own icon
of diversity. Cohabitation:

if we break it down, each iota
— each bit — each it

— is us.

Picking Vegetables

after Sylvia Plath

Stinkhorn, Jew's ear
poison pie, slippery jacks
yellow stainers

we rise before dawn
air damp
from breath

condensed
running in tracks
down shed walls

we gather the dawn spoils
before light
shrivels them

hold up twisted forms
spores carried by wind
to distant homes

and they cannot tell us
what is good
to boil, to fry

what to dry
stew, tincture
for protection

so we leave our prize
heaped on grass
turning to slime

as we're driven
to pull potatoes
from mud

until day-end
and we return
to collapse

as our pile sinks
purge fluid seeping
through cracked earth.

Jane Burn

After the Oilseed

The plough cut hurts in the soil as gulls loitered above,
watching its progress of lines. Barley followed, dropped
through the seed drill, raked beneath uniform tracks.
Moles throw up muddy turrets, break the field's crusted top,
fill up on fresh air. I can feel anger in the earth – it does not
want to be forced from crop to crop with no fallow between.

It wants to bear wildflowers upon its back, be a meadow,
throb with bees – its verges have been joyous.
We humans need to eat, need our bread, our oil, our meat.
I can feel the uncertainties of evening as it waits
for night-time's shadow armies. There is only looking forward
to the day and its passage of time – somewhere, a railway

travels northwest, a ship skirts the wind. Somewhere,
a tree bears a nest of good eggs. I lie, numb in bed while the sky
carries a cow's song, clear and gentle from distant grass.
Somewhere, the unseen story of soldiers, marching.
Somewhere, a quiet line of warheads point towards the sun,
hold stomachs of death tight under their steel skin and bide.

Nearby, poles have been pushed in the naked tract –
where summer's crop was, will be a shoot. At each mark,
a tweeded gun will stand, blast away pheasants for fun
as obedient dogs mouth the dainty back to the man.
I gathered the body of a killed owl – a car had smashed its head
and I cried for pity. It is returning to the ground, its skull

pale behind the dirty feathers of its brow. A heron, far
from the water lands on the furrows, folds into a grey spire,
towers the crows. I feel small. Loneliness whets on my skin
like a blade. I think of the people that ought to be loved
and are not. In this world, there is not enough kindness.
I can feel this slow deflation of dreams.

Flock life

Let me tell you about the rookery,
how I long to feel at home.
Imagine those tall trees,
wired acts of creation
from a scavenger's lifestyle,
how all those loose things,
twigs, leaf, fur and bone
become palaces.

How I long to be amongst my tribe.
Imagine the flock of it all,
the roller-coaster,
big wheel inhalations,
that dream of adrenaline:
down to the arable,
up to the nimbus and squall.

Imagine that taste of happiness,
the belonging, common rook of it all,
symphony of rook-call in your ears.
How I long to beat my wings
in time to a memory of change
and fledge, heartbeat of bird.

Imagine the vastness of sky,
stories of nest and blueness,
all those pecking days, Imagine
all of that going on and on
and never leaving your side.

Judith Cair

Kithurst Meadow

Although the topmost
branches are in
bud, all along the
meadow the ash
trees are dying – not
lost this year perhaps
but soon. Last summer's
lower branches, leafless,
signalled the disease and
now the barren twigs
curl in supplication –
Not yet, not
yet: for one more
spring we'll wear a
crown of leaves – while
below them on tall
stems the cowslips
quiver, cupped to catch
a briefer season.

Contributors

Jill Abram is Director of Malika's Poetry Kitchen, a collective encouraging craft, community and development. She regularly performs her poems in London and occasionally beyond, including Ledbury Poetry Festival, Paris and USA. Publications include *The Rialto, Magma, Under The Radar, The High Window* and *Ink Sweat &Tears.* Jill produces and presents a variety of poetry events and she created and curates the Stablemates reading series.

Audrey Ardern-Jones has been published in *Magma, The Interpreter's House* and many other journals and anthologies. She has been placed or commended in competitions including the Troubadour International Competition. Her first collection ***Doing the Rounds***'will be published by Indigo Dreams later this year. She is Artist in Residence at a London Hospital.

Daphne Astor is an American-born British artist, writer, conservationist and farmer working with UK arts organisations since 1977. In 2016 she founded and curated Poetry in Aldeburgh, has been a trustee at the Poetry School, resident artist at C4RD and Cill Rialaig. She has a BA from New York University, and an MFA from City and Guilds of London Art School.

Clare Best Poetry publications include ***Treasure Ground, Excisions, Breastless, CELL*** and ***Springlines.*** Clare's latest book is a prose memoir, ***The Missing List*** (Linen Press, 2018). A new poetry collection, ***Each Other***, is due from Waterloo Press in summer 2019. Clare recently moved from Sussex to the Suffolk coast. www.clarebest.co.uk

Jemma Borg won the International Ginkgo Ecopoetry Prize in 2018. She also won the RSPB/Rialto Nature and Place Competition in 2017 and is published in magazines including *Poetry Review, Oxford Poetry* and *Plumwood Mountain.* Her first collection is ***The Illuminated World*** (Eyewear, 2014). She has a doctorate in evolutionary genetics and lives in East Sussex.

Carole Bromley lives in York where she is the Stanza rep. and runs poetry surgeries. She has three collections with Smith/Doorstop: the most recent is ***Blast Off!*** (for children). She is published in many anthologies and magazines including *Poetry Review, Poetry News, The Rialto, The North* and *Magma.*

Jane Burn's poems have appeared in a wide range of magazines and anthologies. Her latest collections are ***Fleet*** (Wyrd Harvest Press) and ***One of These Dead Places*** from Culture Matters, where she is an Associate Editor.

Judith Cair lives in West Sussex. She has worked as a potter in rural potteries and as a teacher in an urban middle school. In 2013 a pamphlet of her poetry, ***The Ship's Eye***, was published by Pighog Press. Her work has appeared in several anthologies and she has also written articles on modern interpretations of classical texts.

Claire Collison is one of three winners of the Women Poets' Prize. She was awarded second place in the Resurgence and Hippocrates Prizes, and was shortlisted for the Poetry Business Competition. Artist in Residence at the Women's Art Library, Claire performs her single-breasted life modelling monologue to diverse audiences, including the Ministry of Justice, and a Southend pop-up shop.

Oliver Comins lives and works in West London. His first full-length collection, ***Oak Fish Island***, was published by Templar Poetry in 2018. 'Cornflowers' is from a calendar of 12 poems set in Pitshanger Park, Ealing.

Sarah Doyle is the Pre-Raphaelite Society's Poet in Residence. She is widely placed/published, winning the WoLF poetry competition and Brexit in Poetry 2019, and being highly commended in the Ginkgo Prize for Ecopoetry and in the Forward Prizes 2018. Sarah is co-editor of ***Humanagerie***, a recent anthology from Eibonvale Press, and is currently researching a PhD in meteorological poetry at Birmingham City University.

Claire Dyer's latest novel, ***The Last Day***, is published by The Dome Press. Her previous novels are published by Quercus and her poetry collections are published by Two Rivers Press. She is a regular guest on BBC Radio Berkshire's Radio Reads, teaches creative writing and runs Fresh Eyes, an editorial and critiquing service. Her website is www.clairedyer.com

Suzannah Evans lives in Sheffield and her first poetry collection, ***Near Future***, was published by Nine Arches Press in 2018. She is a previous winner of the Poetry Business Book and Pamphlet Competition and is a Gladstone's Library Writer in Residence for 2019.

Martin Figura's ***Shed*** (Gatehouse Press) and ***Dr Zeeman's Catastrophe Machine*** (Cinnamon Press) were both published in 2016, and a new edition of ***Whistle*** (Cinnamon Press) in 2018. The spoken word show 'Dr Zeeman's Catastrophe Machine' is currently touring. He lives in Norwich with Helen Ivory and sciatica, where he runs the literature event Café Writers. www.martinfigura.co.uk

Charlotte Gann is an editor from Sussex. Poems have appeared in *The Rialto, The North* and *Magma*, and her pamphlet, ***The Long Woman*** (Pighog), was shortlisted for the 2012 Michael Marks Award. Her full collection, ***Noir***, was published in 2016 by HappenStance.

Rebecca Gethin lives on Dartmoor. She had two pamphlets published in 2017 (Cinnamon Press and Three Drops Press). She has been a Hawthornden Fellow and undertook a Residency at Brisons Veor. ***Messages*** was a winner in the Coast to Coast to Coast pamphlet competition. ***Vanishings*** is forthcoming from Palewell Press.

Anuja Ghimire is a native of Kathmandu, Nepal. A Best of the Net and Pushcart nominee, she writes poetry and flash fiction. Her chapbook ***Kathmandu*** is forthcoming from Unsolicited Press in 2020. She lives near Dallas, Texas with her husband and two daughters. She works as a senior publisher/editor in an education-based company.

James Goodman is from Cornwall and lives near London. He works as an environmentalist and futurist for a sustainability charity. His first collection of poems, ***Claytown***, was published by Salt in 2011 and he is working towards a second collection provisionally titled Stone Mountain Fairy Shrimp.

Philip Gross has published some twenty collections of poetry, winning the T.S.Eliot Prize in 2009, and a Cholmondeley Award in 2017. He is a keen collaborator – with artist Valerie Coffin Price on ***A Fold In The River*** (Seren, 2015) and poet Lesley Saunders on ***A Part of the Main*** (Mulfran, 2018). A science-based collection for young people, ***Dark Sky Park***, appeared from Otter-Barry Books, 2018.

Jean Hall lives in London and has held a variety of posts including: Librairie Hachette, London; translator in Paris; teaching English at the Bri-Am Institute, Madrid; Stills Librarian at Warner-Pathé; travel photographer for BT, Denmark. She also competed for GB journalists in international ski competitions. Jean's poems have been published in anthologies, newspapers and magazines. She's currently working on her first collection.

Caroline Hammond lives in London and is a founding member of LetterPress Poets. Her poems have appeared in anthologies and online journals including *Ink, Sweat and Tears*.

Susannah Hart is a London-based poet who has been widely published in magazines and online. She is on the board of Magma Poetry and her collection ***Out of True*** won the 2018 Live Canon First Collection prize.

Jan Heritage (Editor) has a publishing background and was Promotions Manager for Faber and Faber. Her work is widely published in UK magazines and anthologies. She has Masters degrees in American Literature (Sussex) and Creative Writing (Royal Holloway), where she specialised in Ecopoetics.

Robin Houghton's third pamphlet ***All the Relevant Gods*** (2018) was a winner in the Cinnamon Press Pamphlet Competition. Poems appear in many magazines, and awards include the Hamish Canham Prize and the Stanza Poetry Competition. She's written several books on blogging and social media, most recently ***A Guide to Getting Published in UK Poetry Magazines***.

Matt Howard lives in Norwich where he works for the RSPB. His first full collection, ***Gall***, was published by The Rialto in 2018.

Lisa Kelly is Chair of Magma Poetry. Her first collection ***A Map Towards Fluency*** is forthcoming from Carcanet this summer.

Asim Khan lives and works in Birmingham/Coventry. His poetry has appeared and is forthcoming in various online and print publications: asimk.uk@ecoravel

A Chinese-Malaysian living in London, **LKiew** earns her living as an accountant. She holds a MSc in Creative Writing and Literary Studies from Edinburgh University. She was shortlisted for 2017 Primers mentoring and publication scheme and took part in the TOAST mentoring scheme. Her debut pamphlet ***The Unquiet*** was published by Offord Road Books in February 2019.

Fiona Larkin's poems appear in journals and anthologies, such as *Magma*, *The North* and *Under the Radar*, and *Best New British and Irish Poets 2018*. Work was shortlisted in 2018 for the Bridport Prize and the Aesthetica Creative Writing Award. She organises events with Corrupted Poetry, and has a Creative Writing MA from Royal Holloway.

Pippa Little is a Royal Literary Fund Fellow at Newcastle University. Her last book ***Twist*** was shortlisted for The Saltire Prize and she's currently working on her third collection.

Jane Lovell has been widely published in journals and anthologies. She won the Flambard Prize in 2015 and has been shortlisted for several awards including the Basil Bunting Prize, the Robert Graves Prize and Periplum Book Award. Her pamphlets have been published by Against the Grain Press, Night River Wood and Coast to Coast to Coast. ***This Tilting Earth*** is to be published by Seren later in 2019. Jane also writes for *Elementum* Journal.

Roy Marshall's poetry collections are ***Gopagilla*** (2012) ***The Sun Bathers***'(2013) and ***The Great Animator*** (2017). He is currently working on translations of Eugenio Montale to be published in autumn 2019.

Katherine McMahon is a performance poet. As well as writing and performing, she runs participatory projects and events, and has an MA in Creative Writing and Education. She debuted her spoken word show, Fat Kid Running, in 2017. She aims to use poetry to build a more just, more sustainable, kinder world through community and solidarity. You can find her at http://katherinemcmahon.org

Cheryl Moskowitz writes for adults and children. Formerly an actor, she is trained in psychodynamic counselling and dramatherapy. She is an editor at Magma Poetry and on the organising committee for the European Psychoanalytic Film Festival. Together with her musician husband, Alastair Gavin, she co-hosts The All Saints Sessions, a bi-monthly programme of experimental poetry and electronic music in a candlelit church setting in North London. https://www.facebook.com/allsaintssessions/

Katrina Naomi's ***Typhoon Etiquette*** will be published by Verve Poetry Press in April 2019, following an Arts Council-sponsored project in Japan. Her most recent collection is ***The Way the Crocodile Taught Me*** (Seren, 2016). She recently received an Author's Foundation Award from the Society of Authors. Her third full collection is due from Seren in 2020. Katrina lives in Cornwall. www.katrinanaomi.co.uk

Martin Nathan has worked as a labourer, showman, pancake chef, tire technician, and a railway engineer. His writing concentrates on the strangeness of the everyday, and has been published by Tangent Press, HCE and Grist. His novel ***A Place of Safety*** is published by Salt Publishing. He contributed a monologue to the Young Vic's 'My England' project and is currently studying for an MA at Birkbeck College.

Kate Noakes' most recent collection is ***The Filthy Quiet*** (Parthian, 2019). Her website (boomslangpoetry.blogspot.com) is archived by the National Library of Wales. She lives and writes in London.

Michelle Penn's debut pamphlet, ***Self-portrait as a diviner, failing***, won the 2018 Paper Swans Prize. Poems have appeared in *Shearsman, Magma, Butcher's Dog, Popshot* and other journals. Michelle has read her work on Resonance FM and was interviewed in the Radio 4 documentary, 'The Noisy Page.' She recently completed a book-length poem that re-imagines The Tempest as a violent dystopia.

Richard Price's latest collection is ***Moon For Sale*** (Carcanet). He is a lyricist, vocalist, and mauly guitarist for The Loss Adjustors.

EARoberts was a Writing East Midlands mentee 2018/19. She works as an interpreter between British Sign Language and English, with a background in other languages. Her poetry strays between languages and forms, exploring how the visuo-kinaesthetic play of the poem in page/mouth/air interrelates with organic fictions of gender, identity, taxonomy and mythologising.

Sallyanne Rock is a poet and writer living in Worcestershire. Her work appears in various online and print journals, including *Black Country Arts Foundry, Dear Damsels* and *Hypertrophic Literary Magazine*, and features in the anthology *Please Hear What I'm Not Saying*. She has performed her poetry across the Midlands, and currently works with a local young writers group.

Joel Scarfe's poems have featured internationally in magazines and periodicals such as *The Moth, TLS, Gravel, South Bank Poetry, Ambit, The North* and *The London Magazine*. He lives in Bristol with the ceramicist Rebecca Edelmann and their two children.

Penny Sharman is a poet, artist, photographer and therapist. She has an MA in Creative Writing and has been published in magazines and anthologies over the last ten years. Her debut pamphlet ***Fair Ground*** (Yaffle Press) is due out this spring. Penny is inspired by natural landscapes, the cosmos and humanness. She has a surreal approach but grounds her poems with the everyday.

Lesley Sharpe teaches literature and creative writing in London. Most recently her poems have been published in the *Aesthetica* Creative Annual and the Emma Press anthology ***Dragons of the Prime***, shortlisted for The London Magazine, Aesthetica and Bridport prizes, and long listed for Primers 2 (Nine Arches) and Cinnamon Press 2018 Debut Collection Prize. She edits *Heron* for the Katherine Mansfield Society, and is a co-founder of Lodestone Poets.

Paul Stephenson grew up in Cambridge. He took part in the Jerwood/Arvon mentoring scheme in 2013/14 and has published three poetry pamphlets: ***Those People*** (Smith/Doorstop, 2015), ***The Days that Followed Paris*** (HappenStance, 2016) and ***Selfie with Waterlilies*** (Paper Swans Press, 2017). He is co-curating the Poetry in Aldeburgh festival 2019 and interviews poets on his blog at paulstep.com

Matthew Stewart works in the Spanish wine trade and lives between Extremadura and West Sussex. Following two pamphlets with HappenStance Press, both now sold out, his first full collection, ***The Knives of Villalejo***, was published by Eyewear Books in 2017.

Marion Tracy lives in Brighton. She's had a pamphlet, ***Giant in the Doorway*** published with HappenStance, and a first collection ***Dreaming of our Better Selves*** with Vanguard editions.

Lydia Unsworth is the author of two collections of poetry: ***Certain Manoeuvres*** (Knives Forks & Spoons) and ***Nostalgia for Bodies*** (Erbacce), for which she won the 2018 Erbacce Poetry Prize. Recent work can be found in *Ambit, Pank, Litro, Tears in the Fence, Banshee, Ink Sweat & Tears*, and *Sentence: Journal of Prose Poetics*. Based in Manchester/Amsterdam.

Becky Varley-Winter is from the Isle of Wight, and lives in London. Her debut poetry pamphlet, ***Heroines: on the Blue Peninsula,*** is published by V. Press (May 2019). Her poetry has also appeared most recently in Sidekick Books' ***No, Robot, No*** anthology, *Tentacular* magazine, *Lighthouse Literary Journal, Rising*, and *Poems in Which*.

Jane Wilkinson is a landscape architect currently living in Norwich. She has been published in magazines and anthologies including *Ink Sweat & Tears, Envoi* and Emma Press, Live Canon and *The Fenland Reed*. Her competition placings include Cafe Writers, Magma, Rialto and Waltham Forest Poetry. She won the Against the Grain competition (2019).

Hannah Rose Woods is a freelance writer based in London. She recently completed a PhD in Cultural History at the University of Cambridge, and has written for the *New Statesman, the Guardian* and *Elle* magazine. She can be found @hannahrosewoods, tweeting poetic hot takes about ferry contracts, novelty sausages, and the great love of her life, the Bourne Supremacy.

Tim Youngs is the author of the pamphlet, ***Touching Distance*** (Five Leaves, 2017) and co-editor with Sarah Jackson of the anthology, ***In Transit: Poems of Travel*** (The Emma Press, 2018). His poems have appeared in several print and online magazines, including *The Interpreter's House, London Grip, Magma, Salzburg Review* and *Stride*.

Alastair Zangs is based in Norfolk. His days are mostly spent working as an agent for visual artists, outside of this he writes and runs. He is working on his first pamphlet.

www.finishedcreatures.co.uk

Finished Creatures is an independent, no profit magazine, produced by skilled volunteers as a platform for both emerging and experienced poets.

We publish biannually in spring and autumn with two month-long submissions windows in January and July respectively